S0-CMT-603

JOIN THE CLUB

GET INVOLVED IN AN ENVIRONMENTAL CLUB!

BY LISA AMSTUTZ

CAPSTONE PRESS
a capstone imprint

Published by Capstone Press, an imprint of Capstone
1710 Roe Crest Drive, North Mankato, Minnesota 56003
capstonepub.com

Library of Congress Cataloging-in-Publication Data is available on the Library of Congress website.

ISBN: 9781663958839 (hardcover)
ISBN: 9781666320404 (ebook PDF)

Summary: Passionate about the environment? If so, an environmental club might be the right fit for you! Find out what it takes to join an environmental club or start your own, including information on membership, meetings, and activities. Together, you and your fellow members can participate, create, and, most importantly, have fun. Take the plunge, join the club, and get involved!

Editorial Credits
Editor: Ericka Smith; Designer: Sarah Bennett; Media Researcher: Svetlana Zhurkin; Production Specialist: Katy LaVigne

Image Credits
Getty Images: AJ_Watt, 28, Alistair Berg, 24, Ariel Skelley, 14, FatCamera, 29, fstop123, 5, gchutka, 15, GlobalStock, 8, kali9, 20, 22, Leland Bobbe, cover (middle), 27 (bottom), PeopleImages, 16, Plan Shooting 2/Imazins, 27 (top right), Purestock, 13, SDI Productions, 18, SolStock, 7, 10; Shutterstock: Anna Frajtova (doodles), cover background and throughout, Art_Photo, 9, Artgraphixel (blackboard), cover background and throughout, ezpic, 19 (top), govindji, 19 (bottom), Halfpoint, 27 (middle right), irin-k, cover (top and bottom), Jeff Holcombe, 23, Joseph Sohm, 25, K Woodgyer, 27 (cork board), mijatmijatovic, 17, Monkey Business Images, 4, Svitlana Hulko, 21, wavebreakmedia, 27 (left)

Printed and bound in the USA. 4608

TABLE OF CONTENTS

Words in **bold** are in the glossary.

INTRODUCTION

PROTECTING THE PLANET

Do you like watching birds or going to the beach? How about hiking or fishing? Earth is a wonderful place. It is our home. But it faces some big problems, like pollution and **climate change**. And we need planet protectors!

What if you could help Earth *and* hang out with friends at the same time? Good news—you can! Just join an environmental club.

In an environmental club, you can learn more about nature. You can also work with others to help protect the planet.

Can't find a club to join? Don't let that stop you from saving the planet. It's easy to start an environmental club of your own!

CHAPTER 1

ALL ABOUT ENVIRONMENTAL CLUBS

Why should you get involved in an environmental club? There are lots of reasons beyond just helping the environment.

First of all, it's fun! You can get to know people who care about the same things you do. This is a great way to make new friends.

You can learn from others in your club too. Maybe you'll find a new cause that interests you. Or you might find another gardener or butterfly watcher to swap stories with.

You can also work together to improve your environment. Some projects are too big for one person to do alone. Could you clean up a whole beach or plant a grove of trees in a day? Probably not! But a group of people can finish quickly. By working together, you can make a bigger difference.

Finding a Club

So where can you find an environmental club? First, check with your school to see if they have a club you can join. Or ask an adult to help you search for one in your area. You can also check the internet.

Once you find a club, make sure it fits your interests. Then, check when, where, and for how long the club meets. Be sure to show up on time!

What if you can't find a club nearby? You might want to try an online club. These can be fun too. They offer projects that you can do at home. You can share them with other kids in your club.

Starting a Club

If you can't find a club that interests you, you might want to start your own! Maybe you want to focus on decreasing the use of plastics in your community. Or maybe you want to get involved with **activism** addressing climate change. You can start a club to focus on whatever issues you want to tackle—big or small. To get started, you'll need to do a few things.

First, think about what you want your club to look like. Will you focus on big projects? Will you try to learn and teach others?

Next, you will need to find a **sponsor**. This could be a teacher, a caregiver, or another adult. This person will help guide you in getting your club organized. A sponsor can also help you keep things running.

Then, you'll need to find members. Think about who can join. Is there an age limit? You want everyone to feel welcome in your club, but it's okay to set some guidelines.

Talk to your friends and post flyers to let others know about your club. You'll need permission from an adult to hang up flyers at your school and around your community.

Try It! Make a Flyer

Let others know about your club by creating flyers. Make them colorful and eye-catching. You can make your flyers by hand or on the computer. Be sure to include the following information:

- What is the club about?
- Why would someone want to join?
- When and where do club members meet?
- Who should kids contact if they want to join?

CHAPTER 2

DIGGING IN

Now that you have a sponsor and members, it's time to plan your first meeting! This is when you and your new friends will figure out some of the details about your club.

Your First Meeting

At your first meeting, choose a name for your club. The name should give people an idea of what your club is about. Brainstorm ideas and then take a vote.

It's also good to talk about your goals. Some clubs focus on learning about the environment. Others do hands-on projects.

You may want to write a **mission statement**. This tells people the focus of your club. It also helps you keep your goals on track. Do you want to increase the amount of green space in your local area? Do you want to try to protect the environment where a native bird lives? Write it down and make sure your club activities help you accomplish your mission.

Club Meetings

How often will your club meet? Do members need to come to all the meetings? Talk about these questions with your group. Then let new members know the rules when they join. That way, everyone will know what to expect.

Club Costs

Some club activities, like visiting zoos, cost money. To cover your costs, your club might charge **dues**. These are small fees that each member pays every year. Or you might choose to raise money for club activities. You could wash cars or sell donuts. Or you could ask local businesses to donate money.

Club Officers

There are lots of jobs to do in any club. Many clubs have members vote to choose leaders for the year. These leaders are called **officers**. Here are some common roles you might want to include in your environmental club:

- The president plans and runs the meetings. This person asks others to help with jobs.

- The vice president helps the president. This person runs meetings if the president can't be there.

- The treasurer keeps track of the money. This person collects dues and pays for things the club needs. A sponsor may help with these tasks.

- The secretary takes notes at each meeting. This person helps keep track of the club's plans and activities.

Your club could have other leaders too. They might help with other tasks, such as planning projects and welcoming new members.

Meetings

Club meetings can be fun! Everyone has ideas to share. But it's easy for a meeting to get off track. It's good to plan ahead. An **agenda** can help make sure you cover the important things. This is something the club president—or someone else the group chooses—can be responsible for creating.

Try It! Make an Agenda

Create an agenda for your next meeting. Here's an example of what an agenda might look like for your environmental club:

The Cleanup Club

Tuesday, September 28th, 3:00 p.m.

Agenda

3:00 p.m. – Welcome the members.

3:05 p.m. – Take attendance.

3:10 p.m. – Read the notes from the last meeting. Check to see if there are any mistakes.

3:20 p.m. – Discuss plans for the park cleanup on Saturday.

3:55 p.m. – Announce the date and time for the next meeting.

CHAPTER 3

TACKLE A PROJECT

Planning Ahead

Now that your club is organized, it's time for the fun part! Spend some time planning for the year.

There are many kinds of projects an environmental club can do to help the Earth. Think about what will fit your mission. What kind of projects do you want to do? Do you want to tackle one big project, like keeping a park clean? Or would you prefer a few smaller ones, like making reusable napkins and creating bird feeders? It's a good idea to have a mix of activities.

Ask all club members to share some ideas. Make a list. Then let members vote for the projects they like best. Tally the votes, and select one or more projects for the year. Save the remaining ideas for next year.

Project Ideas

Club projects can be indoors or outdoors. Look around your school or home for ideas. What does your neighborhood need? Maybe you could plant a school garden. Maybe you could start a program to **recycle** trash or to plant trees. Picking up litter is a good project for an environmental club too.

Indoor projects are good options when working outside isn't possible, like when the weather is bad. Perhaps your club could make cloth tote bags or sew cloth napkins. These items help keep paper and plastic out of the waste stream. You can share them with family or friends.

Try It! Sew a Cloth Napkin

One simple way to save trees is to make cloth napkins that you can wash and reuse. You'll need help from an adult with ironing and sewing. Ask your sponsor or another trusted adult.

What You Need:

- fabric
- iron
- scissors
- thread
- sewing machine

What You Do:

1. Wash, dry, and iron your fabric.
2. Cut one 18-inch square of fabric for each napkin.
3. Cut a ½-inch triangle off the corners of each square.
4. Fold the fabric over ¼ of an inch on each side of the square toward the back of the fabric. Iron it in place.
5. Fold each side another ¼ of an inch. Iron again.
6. Stitch a straight line along each folded edge.

Learn and Act

Sometimes your club might want to learn something new about helping the environment. You could visit a **nature center** or a zoo. Or you could watch videos about nature. You might want to invite an expert to talk to your group.

Then, put what you learned into action. If you learned about birds, make bird feeders or birdhouses. If you studied pollution, write letters to people in government. Or write an article for the school newspaper. Share what you found out with others in your community. Ask them to help too.

Try It! Plant a Pollinator-Friendly Garden

Pollinators such as bees and butterflies are having a tough time these days. We need these creatures to pollinate many of the fruits and vegetables that we eat. One way you can help is to plant a flower garden at your school or your home.

Start by learning more about the pollinators in your area. What do they like to eat? What plants grow well in your climate?

Once you have that information, clear an area and plant seeds. Keep them watered and weed the area. Put out a shallow dish of water for the insects to drink. Then sit quietly and see which pollinators come to visit!

Go Bigger

Your club might want to join others to tackle a really big project. Welcome friends, parents, and neighbors to help clean up a riverbank or a beach.

Earth Day is celebrated on April 22 every year. Use the day to teach others about the environment. Plan an event for your school or community. Make posters and plan games and crafts. You might even find some new members this way!

CHAPTER 4

SHARE AND CELEBRATE YOUR WORK

Don't be shy about your accomplishments. Spread the word! Let others know about your projects. Show them that kids can make a difference in the environment. You might inspire others to take part in a project too.

One way to do this is to share photos of your work. You could post pictures of your members cleaning up a beach on a bulletin board or send out a newsletter to people who are interested. You might even want to invite a reporter to cover your club delivering a letter about pollution to a member of your local government. (Ask for help from your sponsor or another adult to safely share this information.)

Our Spring Events!

The Cleanup Club

This spring we kept Griffin Park clean. We also learned about the plants that grow in the park and planted some trees.

It's the end of the year. Now there's just one thing left to do—celebrate! Plan a party. Bring snacks and drinks. This might be a good time to try out the napkins you made. Perhaps you could sit by a tree you planted. Or maybe you'll visit a beach you helped clean up.

At the party, thank everyone for their hard work. Talk about the ways your group has helped create a cleaner, greener world. You might even hand out awards to members who went above and beyond to help complete your projects and fulfill your mission.

After you've celebrated your achievements, you can start thinking ahead and planning for next year!

GLOSSARY

activism (AK-tuh-viz-um)—working for social or political change

agenda (uh-JEN-duh)—a list of things that will happen

climate change (KLY-muht CHAYNJ)—a significant change in Earth's climate over a period of time

dues (DOOZ)—money someone pays to be a part of an organization

mission statement (MISH-uhn STAYT-muhnt)—a written record of an organization's purpose

nature center (NAY-chur SENT-ur)—a place where people can learn about the environment

officer (OF-uh-sur)—someone who is in charge of other people

recycle (ree-SYE-kuhl)—to make used items into new products; people can recycle items such as rubber, glass, plastic, and aluminum

sponsor (SPON-sur)—someone who is responsible for something

READ MORE

Andrus, Aubre. *The Plastic Problem: 60 Small Ways to Reduce Waste and Save the Earth*. Oakland, CA: Lonely Planet, 2020.

Raij, Emily. *Kids Can Help the Environment*. North Mankato, MN: Capstone Press, 2021.

Zissu, Alexandra. *Earth Squad: 50 People Who Are Saving the Planet*. Philadelphia: Running Press Book Publishers, 2021.

INTERNET SITES

EarthCapades Kids Club
earthcapades.com/join-the-earthcapades-kids-club

Green Kids Club
greenkidsclub.com/register-to-be-a-green-kid-join-the-club

Kids Eco Club
kidsecoclub.org

INDEX

ABOUT THE AUTHOR

Lisa Amstutz is the author of more than 150 books for children and many magazine articles. Her educational background includes a bachelor's degree in biology and a master's degree in environmental science/ecology. A former outdoor educator, Lisa enjoys sharing her love of nature with children. She lives on a small farm with her family.